Jacques Sawadogo

Recycling of aluminium waste in Burkina Faso

Jacques Sawadogo

Recycling of aluminium waste in Burkina Faso

Smelting, refining and physico-chemical characterization of recovered aluminum for the food industry

ScienciaScripts

Imprint
Any brand names and product names mentioned in this book are subject to trademark, brand or patent protection and are trademarks or registered trademarks of their respective holders. The use of brand names, product names, common names, trade names, product descriptions etc. even without a particular marking in this work is in no way to be construed to mean that such names may be regarded as unrestricted in respect of trademark and brand protection legislation and could thus be used by anyone.

Cover image: www.ingimage.com

This book is a translation from the original published under ISBN 978-3-639-48427-4.

Publisher:
Sciencia Scripts
is a trademark of
Dodo Books Indian Ocean Ltd., member of the OmniScriptum S.R.L Publishing group
str. A.Russo 15, of. 61, Chisinau-2068, Republic of Moldova Europe
Printed at: see last page
ISBN: 978-620-4-11891-8

DEDICATION

To our wife **Mrs. SAWADOGO/TRAORE Géneviève** for her patience in collecting data and producing this document.

To our two children **SAWADOGO Joshua Ange Michel** and **SAWADOGO Marie Deo-Gracias**

To our parents and all those who supported me, please accept my gratitude for all the sacrifices and efforts made for my education!

To our mother Alimata GARBAH

To our mentor in scientific research and teaching Professor Emeritus of Analytical Chemistry Jean Boukari LEGMA

In memory of my father Pateinbibla SAWADOGO

TABLE OF CONTENTS

SUMMARY

Recycled aluminum is obtained from aluminum scrap by the smelters. This waste comes from obsolete parts: old metal sheets, boxes, kitchen utensils, and especially from the destruction of car wrecks. In Burkina Faso, aluminium scrap is used to make kitchen utensils and tools for various uses, using methods that induce impurities during the manufacturing process, notably metals that can be toxic. The primary objective of this work is to analyze this artisanal practice of foundry in Burkina Faso for a better understanding of this method of manufacture, and then to detect the actual presence (or excess content) of these impurities in the objects manufactured. The perspective is to study the methods that could allow the craftsmen founders to respect a certain dosage of the elements of addition which would guarantee a food safety, a better manufacture by moulding of the kitchen utensils and tools of various use.

INTRODUCTION

Aluminium with its very low melting point is very suitable for on-site recycling. Its treatment requires means which are within reach. Its high price since always (enormous energy consumption for extraction) still encourages its recovery. All these factors explain its recycling for so long. The technique used for the manufacture of local kitchen utensils is the foundry. Indeed, the artisanal foundry is a traditional activity in African countries. It is also one of the ways to shape the material in an economical way, because the work of massive metals requires material means that only industrial companies have in Africa. This activity remains the prerogative of the artisanal sector which, despite very limited means, manages to produce quite complex objects. The only materials available are lead, tin, bronzes, aluminum and its alloys. These materials, thanks to their low melting points ***(Baralis et al., 1995),*** can be melted using rudimentary techniques based on the combustion of charcoal, which is widely available even in the most arid regions of Africa. In addition, aluminum and many aluminum-based alloys are increasingly being produced from scrap and used by artisan smelters to make a variety of objects in West and Central African countries. However, many of these objects such as pots, spoons, ladles, pans and knives are used in most of our kitchens ***(Sinou, 1966)***. The manufacture of kitchen utensils is not subject to any precise specifications. However, the products manufactured must be resistant to fire and to the mechanical stresses generated by their handling, while having an average life of about ten years. In Burkina Faso, no serious study has yet been done in this area, although it could have a negative impact on the health of the population. This is why our work aims to study and analyze this artisanal practice of casting in Burkina Faso for a better understanding of this method of manufacture.

In this valuation document we present :

S the results of laboratory analyses of the contents of the various components of samples obtained in four localities of the city of Ouagadougou (Burkina Faso);

A comparison ***(Table 1)*** *of* these results with extracts from the French standard NF 601 of July 2004.

However, it is essential to present the different stages of manufacturing of this material before reaching the users in order to be able to circumscribe in due time the problems and the improvements to be recommended.

I. METHODOLOGY

The methodology of the technique is carried out in several steps starting with the recovery of the aluminium waste.

1. Aluminium scrap recovery

The first step in waste recovery is waste collection. In Burkina Faso, the collection of aluminum waste is done in an artisanal way, mainly at public dumps, municipal impounds and car repair garages. The most frequently encountered items in this aluminum waste ***(Figure 1)*** *are*: packaging, transport, household materials, food industry, engine heads, car rims, cylinder housings, gearbox or clutch housings, water pump bodies, iron soles, pistons, electrical conductors, beverage cans, scrap from stamping factories, printing plates, profiles and Yamaha cylinders, etc. Real collection networks (sometimes transnational) are organized upstream of the production units and the quantities of aluminum recovered are very large, judging by the number of objects produced and available for sale on the markets.

Figure 1: Examples of aluminum alloy waste (Sawadogo et al., 2014).

A thorough sweep of the neighborhoods of Ouagadougou revealed four (4) more or less important workshops for the manufacture of recovered aluminum objects. The most important ones are grouped and concentrated in the Goughin district (Goughin I and Goughin II), the Zone I district and the Zone non lotie district of sector 30. The samples were named as follows: Sample #1 (Zone 1); Sample #2 (Goughin I), Alloy Sample #3 (Goughin II) and Alloy Sample #4 (Unallocated Zone).

The price of these wastes does not depend on their chemical composition, but rather on the supply and demand in the market. These recovered objects are currently sold to artisans at a price between 500 CFA ***(about 90 cents)*** and 700 F CFA ***(about one euro)*** per kilogram in Ouagadougou. Although the precise metallurgical knowledge of the alloys used in the manufacture of cooking pots escapes the artisans as a whole, they have a very pragmatic perception of this raw material.

The artisanal smelters of Burkina Faso have acquired empirical knowledge of aluminum metallurgy that allows them to sort the different types of "white metal" and to identify the many forms of alloys present in the metals they collect. They are then able to mix and refine these different types of aluminum to obtain the ideal alloy for their various products. The melting point of aluminum is 640°C. But in order to be cast, it must be heated to 880°C, at which temperature it becomes completely liquid. This temperature can be reached in rudimentary furnaces, heated with charcoal. These are of various sizes and shapes and operate on a similar principle. The main part is formed by a kind of corolla made of a mixture of clay and molding sand as shown in ***Figure 2***. This enclosure hardens on contact with the fire, but nevertheless remains brittle and must be regularly repaired. Charcoal is placed in the bottom of the furnace (**5**) and a nozzle (**2**) fed by a blower (**1**) brings air to force the combustion. A channel (**4**) placed slightly below allows the evacuation of the fumes, but also the recovery of the aluminium when the crucible is accidentally pierced during the melting process ***(Facy et al., 1983)***.

Figure 2: Oven: general aspect and cross section *(Agier, 1981)*

2. Aluminum recycling - Melting of the charge

2.1. Description of the process

Once the charge is made up, the craftsman prepares for the second stage of his recovery process: melting. The aluminum scrap does not undergo any prior preparation (cleaning, etc.). They are placed in a container that serves as a crucible and then placed in the artisanal furnace. This means that plastics, ferrous and non-ferrous metals can become trapped in the scrap and "pollute" the molten aluminum. Not to mention the traces of oil, grease, paint and even soil that are brought to high temperatures.

Melting is carried out in a furnace dug into the ground to an average depth of 30 to 50 cm, equipped to receive the crucible ***(Figure 3)***.

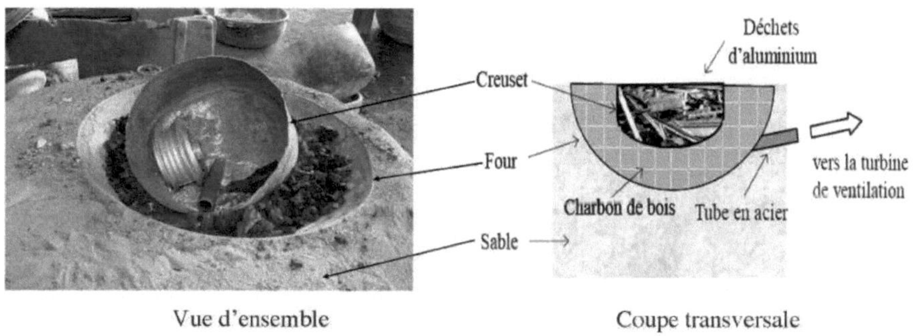

Figure 3: Artisanal oven (Ndiaye, 2006)

The furnace is ventilated through a steel tube that opens under the crucible. The other end of this tube is connected to a manually operated turbine ***(Figure 4)***, which is responsible for propelling the air necessary to activate the combustion of the charcoal.

Figure 4: Ventilation turbine of the artisanal oven *(**Sawadogo, 2015**)*.

This turbine is a recovery material from industrial ventilation systems or from oil combustion devices on industrial heating systems. Each smelter has an average of three crucibles of different sizes which are used according to the quantity of aluminium to be smelted. During the melting process, the crucible is covered with a similar plate, thus limiting the energy loss by radiation.

The melting is achieved thanks to the energy provided by the combustion of charcoal, which remains the most used domestic fuel in Africa, especially in the countryside, because it is available and inexpensive. Indeed, this melting is long, depending on the quantity of material contained in the charge: for example, to melt about twenty kilos of aluminum, 1h30 to 2 hours of heating are necessary. During this operation, the furnace is regularly replenished with charcoal. The crucible is then moved to allow a good distribution of charcoal in the furnace. In practice, it happens that some pieces of charcoal fall into the molten metal where they burn and turn into ashes that mix with the slag. When the molten metal reaches a temperature of between 650°C and 800°C, which can only be judged by its color, the craftsman "scours" the molten metal bath. He removes the crust of alumina and slag that has formed on the surface ***(Figure 5a)*** and removes the pieces of iron or steel mixed with the raw material that cannot be melted at the temperatures reached during the melting of the aluminum ***(Figure 5b)***.

(a) (b)

Figure 5: Operations carried out at the end of the melting process: (a) removal of the alumina crust
and slag; (b) residues of unfused iron or steel

2.2. Moulding

The mold is the device in which the craftsman will make the imprint of the piece he wants to reproduce. Molds can be divided into two groups: non-permanent molds and permanent molds

2.3. Non-permanent moulds

They are destroyed after the metal has solidified to extract the manufactured piece. This is the case of the molds made with silico-clayey sand by the Burkinabe craftsmen who make the pots.

2.4. Permanent molds or indestructible molds

They are made of metal and this technique allows to reproduce a large number of copies of the part to be duplicated without having to build a new mold. But this mold has a disadvantage, because it allows to produce only one and the same object unlike the sand mold. Most of the molds used for the manufacture of kitchen utensils in the environment are non-permanent molds incorporating destructible cores. The cores allow for the molding of the interior shapes of these utensils. In the Ouagadougou workshops, the technique used is called "silicoclay sand". It consists of using a wet sand-clay mixture that has the property of precisely following the contours of the model, of hardening when "squeezed" ***(Youssouf, 1996)*** and thus of perfectly preserving the imprint of the object. Moreover, the "silicoclay sand" is a refractory material, which allows it to resist contact with the molten metal without deforming. The complexity of the molding process is linked to a characteristic of the object to be reproduced, which is called the "skin" ***(Facy and Pompidou, 1983)***. Indeed, the more easily the object is "stripped", the simpler the device will be to elaborate.

2.5. Materials used to make local cooking utensils In order to contain the sand, the craftsman uses frames called chassis ***(Figure 6)*** of different sizes, usually made of wood, pestles and wooden sticks used to pack the material

Figure 6: Example of a frame (Ndiaye, 2006)

The dimensions of the frames *(Table 3)* are adapted to the size of the local cooking utensils to be manufactured

Table 1: Dimensions of the frames for the manufacture of local cookware

Frame dimensions L x T x H (mm)	*Names of kitchen utensils (Kg)*
510 x 510 x 440	30
450 x 465 x 350	20
440 x 430 x 350	15
350 x 340 x 225	10 and 7
330 x 320 x 240	5 and 4
300 x 305 x 190	3
330 x 335 x 225	2 and 1.5
220 x 230 x 190	1; 0.5 and 0.25

The making of a mold requires great manual skill and an excellent knowledge of the molding material. The humidity of the sand, its clay content, the way it is "tightened" progressively and regularly around the model determines the success and quality of the casting and the object to be produced. The cross-section of the mold ***(Figure 7)*** allows us to better understand the principle of the device. We can see in particular the impression left by the model in the sand (**6**) and the pouring orifice (**4**) through which the liquid metal will penetrate into the mold before filling the whole impression.

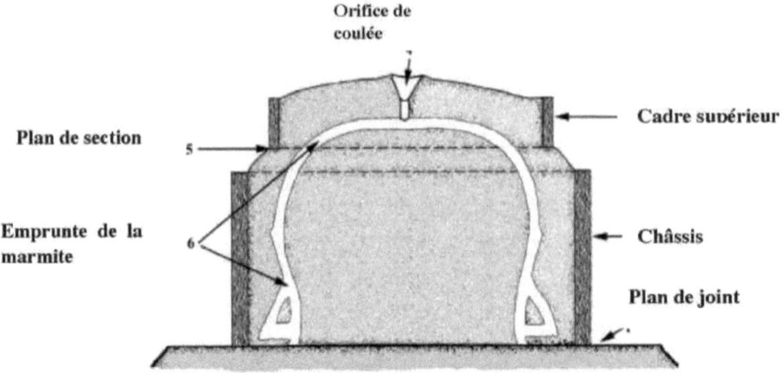

Figure 7: Cross-section of the moulding device

a) Preparation and filling of the mould

Among the different casting techniques known today, sand casting with permanent pattern is the one used by the craftsmen-founders of Ouagadougou. These aluminum artisans cast their objects in sand molds, which must be rebuilt after each operation. For the casting of an object, the inner mold is obtained by compacting very fine and wet sand on the inside. For the outside, sand is also applied to the object in two or more parts. After compacting the sand, the model is removed. The entire casting surface, which will be in contact with the molten aluminum, is covered with a layer of lime. The joining surfaces of the two or three parts of the mold are coated with a thin layer of dry laterite powder to help separate the elements. Some artisans draw illustrations on the inside surface of the mold with the back of a spoon to make the pot look more attractive. After removing the metal model and reassembling the mold parts, the aluminum is poured into a prepared hole.

b) Casting and unmoulding

The third step in the process is casting and demolding. The first step is to fill the impression left by the object with molten metal. Using long-handled blacksmith's tongs, the caster grabs the crucible containing the aluminum and carries it at arm's length to the mold. The liquid metal, which is now at a temperature of around 850°C, is poured into a small funnel in one go. An assistant holds the different parts of the mold to prevent the gas pressure produced by the contact of the metal with the wet sand from dislocating the device. A perfect control of the gestures is essential to guarantee a good filling of the mould, but also to avoid overflowing, which could lead to serious burns. Once the metal is poured, the caster ensures its progressive cooling by spraying the mold with water. Finally, he extracts the object by destroying the molding device.

c) The different stages of the manufacture of kitchen utensils in Burkina Faso

The description of the operating chain has made it possible to highlight the nature and diversity of the knowledge and skills required in the practice of aluminum casting. Regardless of the location of the observation, the most striking aspect is the terribly homogeneous nature of the manufacturing processes. Of course, the objects produced may vary in shape, decorative patterns or degree of finish. But the method used to produce them remains invariable. The homogeneity of the technique thus appears to be a central characteristic of the distribution of the aluminum foundry. This is in stark contrast to other techniques studied in Africa, which show procedural variations that are sometimes very important. This is notably the case for iron metallurgy or pottery. In addition to the fact that aluminum forming techniques are relatively limited, this technical homogeneity is due to a mode of transmission of knowledge by apprenticeship, where practice occupies a fundamental place.

The operating chain for the implementation of local kitchen utensils is as follows:

1. The pot has no "skin", so it is necessary to use a two-part model that will keep the internal print intact *(**Figure 8**);*

Figure 9 : placement (assembly) of the model

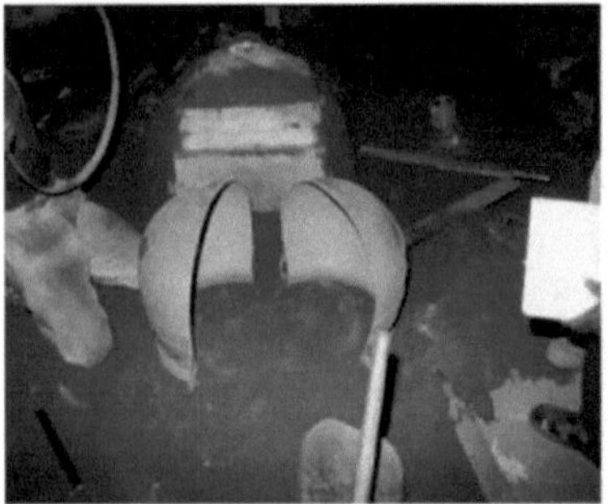

Figure 8: The two parts of the model

2. the two parts of the model are assembled and strapped with a flexible metal band whose length is adapted to the circumference of the pot.

The tension of the strapping is ensured by the insertion of a wooden wedge ***(Figure 9) ;***

3. After being placed and immobilized in the center of a mound of loose sand, the interior of the model is filled with several layers of sand that are successively compacted (tightened) with a wooden pestle. The last layer of sand is rammed with a bat and forms a perfectly flat and hard surface at the opening of the model ***(Figure 10).***

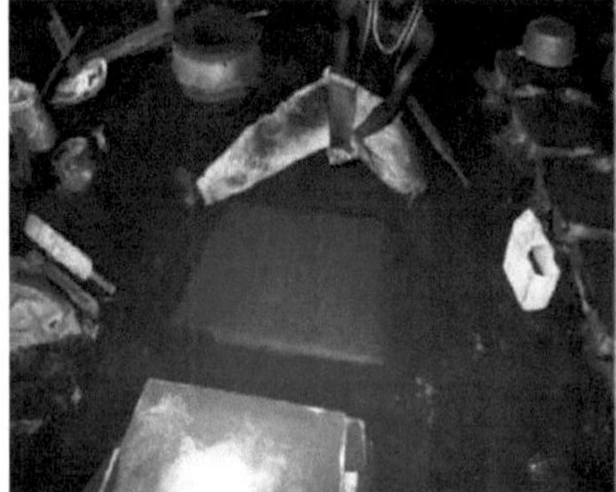

Figure 10: Preparation of the impression **Figure 11**: Preparation of the base of the internal model of the moulding device

4. the elaboration of a flat base on which the moulding device will be built by

"straightening" (which consists in giving a flat shape) a layer of sand with a ruler (long wooden bat). The sand is rammed with the bat ***(Figure 11).*** The base thus formed is hardened by the grooming, perfectly flat and square in shape. The base is sprinkled with dry sand to create a horizontal sectional plane ***(Paul et al., 2006).***

5. The model is tilted, turned over and placed in the center of the base. The strapping is removed and two metal plates are inserted into the handles of the pot ***{Figure 12).*** The strapping is removed and two metal plates are inserted into the handles of the pot (they will form the vertical sectional plane of the 2 lower half frames) ***(Tamari, 1991).***

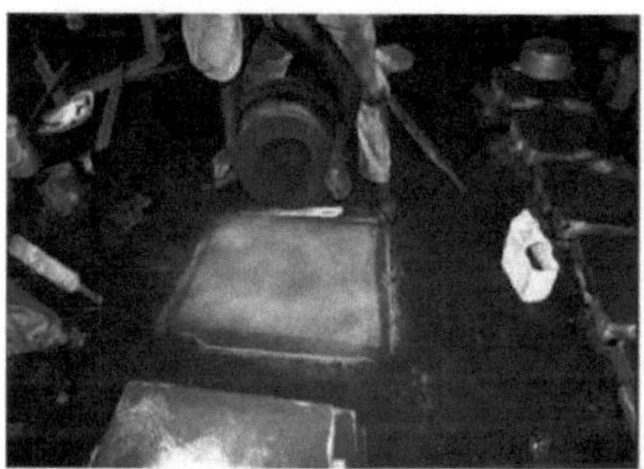

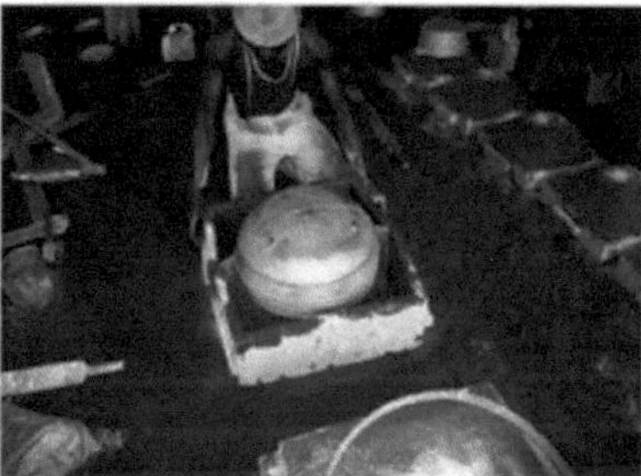

Figure 12: Positioning of the model **Figure 13**: Placement of the half-frames On the lower base

6. the 2 wooden half-frames are moistened, placed on either side of the model and secured with a ligature ***(Figure 13).***
7. The frame thus formed is progressively filled with layers of sand "squeezed" with the help of a pestle and the wooden bat ***(Figure 14).***
8. a new sectional plane is made with a spatula (around the bottom of the model) and sprinkled with sand ***{Figure 15).***

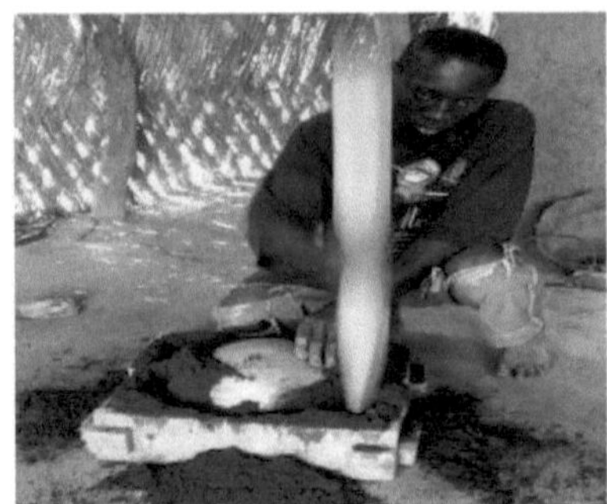

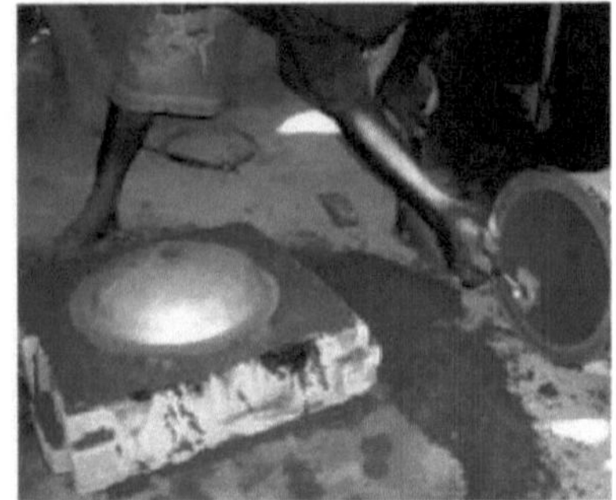

Figure 14: Filling in the lower frame **Figure 15:** Drawing up the upper section plan

9. After being moistened, the upper frame is laid, filled with layers of sand that are successively squeezed with a pestle. A sand dome is formed above the bottom of the model and rammed with the bat *(**Figure 16**).*

10. a small hopper is dug at the top of the dome. Using a metal tube driven to the contact of the model, a core of sand is extracted, thus clearing the pouring conduit *(**Figure 17**).*

Figure 16: Filling the top frame

Figure 17: Preparation of the pouring hole

11. At the base of the upper frame and on two opposite sides, the sand is smoothed with a spatula and covered with a thin layer of talc. The talcum powdered sand is then "marked" in the extension of pre-existing grooves on the frame in order to determine the exact position of the upper frame ***(Figure 18).***

12. The upper frame is removed. Imperfections are corrected with a metal cup. The impression is then trowelled ***(Figure 19).***

Figure 18: Locating the position of the top frame

Figure 19: Installing the upper part of the mold

13. The lower frame ligatures are undone. A separation of the vertical parting line of the lower frame is obtained by slightly rotating a metal blade between the two parts of the model. One of the lower half-frames is removed. The "half model" is extracted after being lightly impacted in order to promote its lift-off. Corrections are made to the impression (mainly to the handles). The impression is trowelled ***(Figure 20);***

Figure 20: Disassembly of the half-frames **Figure 21**: Complete disassembly of the lower mold

2. The same operations are performed for the other lower frame. The metal plates forming the vertical section plane are removed. Small defects in the impression are corrected with a metal spatula. The impression is then extensively trowelled ***(Figure 21) (Sawadogo, 2015)*** .

3. The first half of the lower frame is put back in place. The positioning of this frame ***(Figure 22)*** is crucial. It must be ensured that the same space is observed everywhere between the inner and outer cavities of the model. Any misalignment will result in poor distribution of the molten metal in the cavity space during casting, with the result that holes will appear in the object produced.

4. The second part of the bottom frame is replaced and should fit perfectly with the first ***(Figure 23)***.

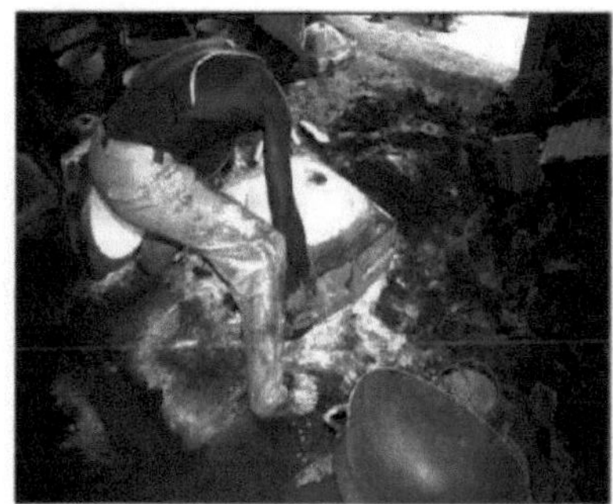

Figure 22: Installation of a half frame

Figure 23: Installation of the second bottom frame Lower

17. The top frame is repositioned and adjusted using the marks made earlier. The pouring hole is then covered with a cover to prevent sand or debris from entering the mold cavity. Sand is then piled up along the 2 walls of the lower frame parallel to the vertical sectional plane to consolidate the molding device ***(Figure 24a).***

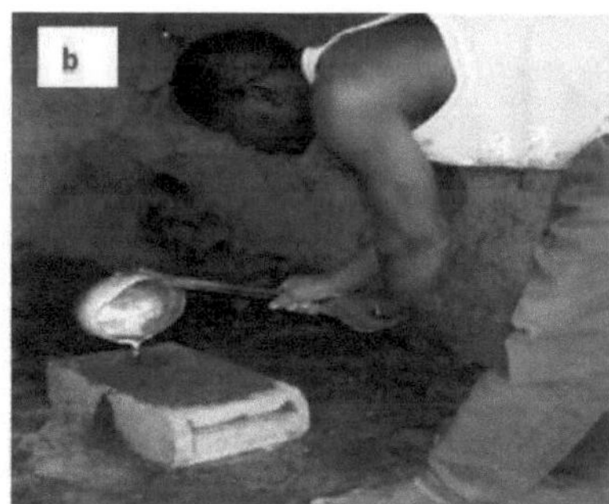

Figure 24: a) installation of the upper frame, b) casting

18. The liquid metal is poured into the mold using a crucible. An assistant holds the molding device with pieces of wood so that it does not break apart under the thrust of the gases caused by the meeting of the molten metal and the slightly moist sand ***(Figures 24b and 25a).***

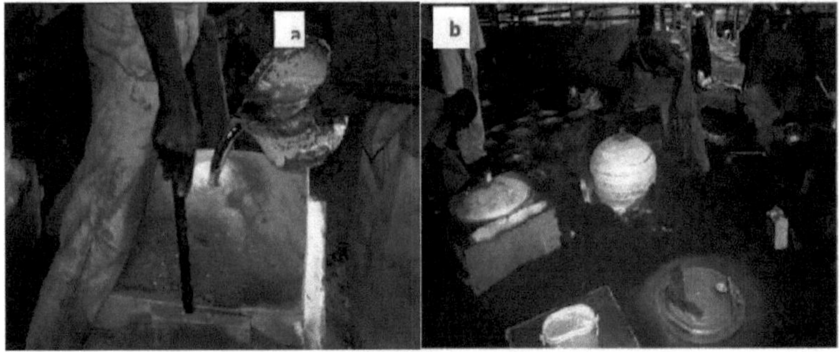

Figure 25: a) Casting; b) Demolding

5. Once the metal has solidified and cooled, the mold is destroyed and the pot is removed ***(Figures 25b and 26) (Ndiaye et al., 2009).***

6. The pot is cleared of its feeder (the growth left by the filling hole) by sawing and deburred by filing ***(Figure 27)***. This stage of the production process is called finishing.

Figure 26: Removal from the mould: stripping the interior and de-sheathing

Figure 27: Tool used to surface pans

Finishing is the final step in production. Once the feeder is sawn (**Figure 28**), the excess castings are deburred and the surface of the object is scraped and polished to make it smooth and uniform in appearance. The degree of finishing varies greatly from observation site to observation site and can sometimes be very rough. In some cases, small holes caused by casting defects are also filled. Since aluminum cannot be welded without specialized tools, the craftsman must then proceed with a form of riveting using a piece of flexible aluminum that will be caulked into the hole to be sealed. This type of repair must be as discreet as possible, because it is a defect that affects the longevity of the object and that buyers try to spot. At the end of these last operations, the object is ready to be marketed ***(Elliot et al., 2004).*** In other words, after demolding, the objects enter a finishing stage before being marketed. The sprue is separated from the kettle by sawing ***(Figure 28)***. This will be recycled in the following castings. The objects are then given to a worker to deburr and polish the surfaces to make them more functional and attractive. The outer surface of the pot is filed down, while the inner surface is scraped with a curved, sharpened steel blade. This makes the inside of the pot easier to clean after use.

Figure 28: Sawing of the sprue

The objects are then exhibited, sometimes at the smelter, but most of the time they are taken to a dealer ***(Figure 29)*** *to* await a buyer.

Figure 29: Display of kitchen utensils (finished products) at the artisan's home

II. LIMITATION OF THE RECYCLING TECHNIQUE

A comparison of aluminum recycling with other metallic materials highlights the low ranking of aluminum, with an average recycling rate of 37%, compared to 50% higher for steel, copper and lead. This result is surprising because aluminum recycling is continuous and uses very little energy.

Recycling is done exclusively by fusion and the application of fundamental laws of chemistry that highlight the limitations to eternal recycling:

- there are losses of aluminum, irreversible when aluminum is transformed into alumina, reversible when aluminum is dragged by alumina. The alumina-aluminum mixture being the dross, it leads to the increase of the concentration of other elements.
- The artisanal purification of molten aluminum is very basic: it is limited to simply skimming with a perforated ladle the varnish and other dirt floating on top of the molten aluminum.

Artisanal recycling is done in a single step as opposed to the minimum four steps recommended in modern recycling industries. Sand molds leave a rough imprint on the surface of the molded materials. This roughness will play a detrimental role in the resistance to aqueous corrosion, as it increases the reactivity at the joints. The hollows can be the seat of corrosion by concentration difference. On the other hand, simple gravity casting induces the formation of micropores on the surface of the aluminum due to a too fast cooling or to a problem of air evacuation which often leads to an incomplete filling of the mold.

Other technical problems include: those who keep large quantities of aluminium scrap at home in case they need a new pot cause problems for the smelters: on the one hand, part of the aluminium is tied up in the process for a while. On the other hand, it is increasingly common for customers to come with their own raw material and order a specific product at very short notice. The craftsman who normally makes a series of castings every few days is therefore forced to light the fire and do all the work involved in casting aluminum for one object only. Producing one object at a time increases the consumption of coal and labor in general.

Some smelters felt that the amount of aluminum in the city of Ouagadougou was decreasing, while others felt that aluminum has always been scarce and will remain so. The lack of aluminum forces all artisan smelters to take long breaks from work. The lack of coal in the city has further increased the unemployment phases. In order to work, you need aluminium, coal and a crucible.

III. CHEMICAL ANALYSIS OF RECOVERED ALUMINUM ALLOYS

The contents of the constituent elements in the recovered aluminum alloys were identified and quantified by several of the techniques:

ICP-OES technique: Each sample is dissolved by acid attack in a mixture of hydrochloric acid - milli-Q water (50 : 50). The analyses are performed on 10 ml samples of the solution containing the dissolved samples. For the determination of major elements, the initial solution was diluted to obtain trace concentrations to remain in the Beer-Lambert range ***(Müller et al., 1993; Woodson, 1998)*** on the spectrometer;

The X-ray fluorescence technique allows a qualitative and semi-quantitative analysis. The device used is a spectrometer type SRS 3000;

The DRX technique allows the identification of the nature of the phases present. The diffractometer used is a Brücker SIEMENS D5000 in mode 20. These different techniques allowed us to analyze 4 samples of various origins.

IV. RESULTS

4.1. Presentation

Each sample was analyzed four times and the result presented is the arithmetic mean of the four measurements. Tables 1 and 2 present the results of the analyses on the samples cast at the foundrymen's.

4.2. Accuracy of results

The different devices used allowed the determination of a maximum of 18 elements in a sample, but some elements were undetectable such as: Cadmium (Tr = 321 °C), Mercury (Tr = 357 °C) and Antimony (Tr = 630.5 °C) whose melting temperature is lower than that of aluminum. The accuracy of the XRF technique used is very dependent on the surface condition of the samples.

4.3. Interpretation of results

The results of the X-ray fluorescence spectrometry analysis of the selected samples have been grouped in Tables 2 and 3 :

a) wide variety of chemical composition of alloys

This is a consequence of the choices made by the smelters to constitute the melting charges. This characteristic testifies to the extreme diversity of the quality and quantity of the materials composing the aluminium scrap deposit, and to a supply at the smelter which is also diversified and variable over time.

b) Silicon is the main additive element in these alloys

Silicon is one of the major constituents of aluminum alloys for foundry use, as it gives with 87.3% of aluminum a eutectic mixture melting at 577°C. It thus lowers the melting temperature of the alloys and improves their castability **(Sawadogo et al., 2014)** and consequently limits the critical risks of cracks in the metal. Aluminum-silicon alloys are sometimes called silumins. The silicon content of homemade aluminum alloys varies from sample to sample (4.27% to 12.7%). They give them a hypoeutectic composition in the Al-Si binary diagram which respects the maximum value of 11.7% given by the NF EN 601 standard of July 2004. The origin of this element is certainly linked to the use in very important proportions of alloys resulting from the automobile industry (engines, car treads, cylinder casings, gearbox or clutch casings, beer cans...).

c) High copper and zinc content

The copper contents in samples No. 2; No. 3; and No. 4 are out of the norm and the zinc contents in all four samples are above the NF EN 601 standard of July 2004.

d) Other elements

Other elements such as nickel, magnesium, titanium, chromium, iron, manganese, zirconium and lead are present in small quantities in the four samples. In fact, all the samples have chloride contents clearly higher than those authorized by the French standard, due to the load from which they were obtained ***(Hubulot, 2001)***. The iron and manganese contents of the alloys comply with the standard ***(Mohamed, 2008).*** This confirms what we said earlier that given the absence of pre-established rules for the preparation of the charge of fusion and the diversity of aluminum waste, each alloy is unique. The steel crucible where the molten aluminum remains at a high temperature (over 650°C) for more than an hour is also a source of contamination. The iron contents of the alloys being almost the same, this suggests a common source of contamination. We can then consider two types of contamination of aluminum by iron:

- Aluminum alloys are sometimes combined by construction (composite material) with pieces of steel which, when they are not easy to remove, are put in the charge with the aluminum and removed once the latter has melted. The liquid aluminum being very corrosive towards steel ***(Wang and al., 2003)***, an enrichment by contamination of the liquid metal bath by iron takes place;
- the steel crucible where the molten aluminum remains at a high temperature (over 650°C) for over an hour is also a source of contamination. In addition, the use of tools, also made of steel, such as skimmers, ladles, etc., is a source of contamination.

Lead is not a "natural" constituent of aluminum alloys. It could only be introduced into the fusion charge accidentally. Indeed, lead is not miscible with almost all alloying elements at room temperature, except for tin with which it forms a solid solution containing 98.1% by mass of tin at 20°C. Lead separates by density difference and solidifies last (its melting point 327°C is lower than that of pure aluminum, 660°C). It is then easy to detach it from the part, because it does not adhere to the alloy. If the imprint it leaves causes too great a defect, the product is remelted. However, small lead particles can still remain in the matrix, as we will have the opportunity to show by chemical analysis of the microstructure.

For better use of aluminum scrap, to improve the food quality of kitchenware alloys, which is strongly related to the chemical composition of alloys, we can conclude that :

- the craftsmen must make a judicious sorting, after the collection to select the most suitable waste to enter the confection of the artisanal pots;
- Alloys for kitchen utensils (artisanal kettles) must have good castability to be shaped by casting with sand molds (alloys with high silicon content).
- for a good castability (good shaping in the foundry), alloys from the food industry (can...) and printing plates are preferred, which can be mixed with others from the automotive industry (casings, cylinder heads...) or from the building industry (window frames...).

Table 2: Chemical composition (XRF) in % by mass of samples collected from smelters in Burkina Faso and maximum contents allowed by the NF EN 601 standard (July 2004).

Elements / *samples*	*Al*	*If*	*Fe*	*Cu*	*Zn*	*Mg*	*Cl*	*Cr*	*Mn*	*Pb*	*Nor*	*Ti*	*Zr*	*P*	*S*	*K*	*It*	*Co*
Sample 1	82,3 ±0,33	12,7 ±0,24	0,7 ±0,02	1,1 ±0,02	1,27 ±0,15	0,48 ±0,03	0,19 ±0,03	0,021 ±0,01	0,204 ±0,001	0, *0056 ±0,0 06*	0,31 ±0,01	0,04±0, 01	0,012 ±0,001	0,22 ±0,04	022 ±0,03	0,1 ±0,01	0,079 ±0,012	0,01 7 ±0,0 04
Sample 2	90,3 ±0,34	6,2 ±0,18	0,76 ±0,02	0,85 ±0,01	0,676 ±0,011	0,17 ±0,03	0,14 ±0,03	0,034 ±0,01	0,11 ±0,01	0,03 ±0,0 04	0,061 ±0,004	/	0,011 ±0,001	0,19 ±0,04	0,25 ±0,03	0,079 ±0,03	0,072 ±0,029	/
Sample 3	93,2 ±0,34	4,3 ±0,15	0,6 ±0,02	0,54 ±0,01	0,311 ±0,01	0,24 =0,03	0,13 ±0,03	/	0,055 ±0,01	0,02 ±0,0 04	0,045 ±0,004	0,023 ±0,01	/	0,14 ±0,04	0,2 ± 0,03	0,08 ±0,03	0,076 ±0,011	/
Sample 4	91,8 ±0,34	5,4 ±0,17	0,5 ±0,02	0,73 ±0,01	0,404 ±0,01	0,24 =0,03	O,15 ±0,028	0,02 ±0,01	0,07 ±0,01	0,03 1 ±0,0 04	0,056 ±0,004	0,022 ±0,01	/	0,24 ±0,04	0,23 ±0,03	0,069 ±0,029	0,078 ±0,012	/
Witness	96,8 ±0,34	2,1 ±0,11	0,2 ±0,01	O,01 ±0,01	0,018 ±0,02	<0,01	0,21 ±0,03	<0,005	<0,005	<0,0 02	0,011	<0,01	/	0,18 ±0,04	0,19 ±0,028	0,075 ±0,029	0,14 ±0,015	/
Standard NF 601 of July 2004	remains	13,5	2	0,6	0,25	1[illegible]	<0,05	0,35	4	<0,0 5	3	0,3	0,3	/	/	/	/	/

NF: French Standard

Table 3: Chemical composition given by ICP in % by weight of samples collected from smelters in Burkina Faso.

Elements analyzed **Samples**	**Al**	**Cu**	**Fe**	**Mg**	**Mn**	**If**	**Zn**	**Ti**	**Pb**
Sample 1	81,92 ± 0,201	1,026 ± 0,04	0,626 ± 0,016	0,437 ± 0,02	0,196 ± 0,0036	12,537± 0,082	1,247 ± 0,012	0,043 ± 0,001	0,052 ± 0,002
Sample 2	90,56 ± 0,211	0,827 ± 0,038	0,808 ± 0,016	0,171 ± 0,01	0,122 ± 0,0052	6,516 ± 0, 068	0,596 ±0,018	0,126 ± 0,007	0,024 ± 0,004
Sample 3	93,20 + 0,230	0,559 + 0,041	0,590 + 0,021	0,273 + 0,01	0,052 + 0,0043	4,661 ± 0,057	0,291 + 0,018	0,025 + 0,008	0,022 + 0,004
Sample 4	91,92 + 0,208	0,726 + 0,044	0,480 + 0,024	0,248 + 0,02	0,069 + 0,0047	5,506 ± 0,072	0,415 + 0,025	0,023 + 0,006	0,028 + 0,005

4.4. Analysis of the different phases by X-ray Diffraction

a) Control sample

In this control sample (**Figure 30**), the X-ray diffraction spectrum reveals essentially in the matrix, solid aluminum AlS. Indeed, in this spectrum no foreign phase is revealed. For this sample, the most intense peak observed corresponds to the preferential orientation of the (200) plane. The peaks (111), (220) and (311) observed are of medium intensity.

b) Alloy sample n° 1 (castle Zone 1)

The spectrum of **Figure 30a** shows that the structure of sample n°1 is composed of a solid solution of aluminium as a matrix in which we find essentially as a secondary phase silicon, precipitates of CuZn5 type accompanied probably by particles of Al2Cu in trace form. The literature states in fact, that copper alloyed with aluminum in proportions lower than 55% in mass gives rise to the compound Al2Cu that we find associated with aluminum in a eutectic mixture ***(Baudry, 2007)***. The existence of a very intense peak at 38.47° plane (111) and peaks of average intensity at 44.74°; 56.23° and 65.14° respectively planes (111); (200) and (220) proves that aluminium is the dominant phase of our sample. The presence of foreign phases (silicon and CuZn5) is confirmed by the appearance of very low intensity peaks characterizing their compositions at: 28.443°; 47.30°; 56.12° and 69.13° for silicon and at 37.77°; 41.99°; 43.56°; and 57.88° for CuZn5.

c) Alloy sample #2 (Goughin sector 7), alloy sample #3 (Goughin market) and alloy sample #4 (Undeveloped area sector 50)

In the spectra (c and e) of the Figure concerning samples 2, 3 and 4, the diffractograms reveal mainly the solid aluminum matrix, as the (plane) peaks (111); (200); (220); (311)) of the ASTM (aluminum identity card) of the control aluminum are present. Contrary to the other spectra, whose intense peak is located at 38.47° plane

(111) that of the Goughin II alloy is located at 65.23° plane (220). We note, the presence of small silicon peaks on all the diffraction spectra. However, very small peaks seem to be present on the diffractogram of figure 30c, this phenomenon reflects the presence of foreign phases (Cu, Cu_3Al_2, MnO_2 and $Fe_3Al_2Si_3$) resulting from the rearrangement of atoms in the primitive lattice. The existence of these different phases in this alloy explains the shift of the intense peak to the right.

Note: for copper, it should be noted that it is composed of two very close wavelengths (Ka1 = 1.54056 Â and KB = 1.54439 Â) ***(Shabestari, 2004)***. This explains why, on the diffraction spectra (of the kitchenware samples), we observe a kind of splitting of diffraction peaks. To avoid this, we have exploited for our results the data from the Ka line which has a higher peak intensity.

All the different diffractograms show a preferential orientation or texture at least for the Chateau Zone I alloys, Goughin market, and the unallocated area of sector 50 along the (111) plane and for the Goughin II alloy along the (220) plane.

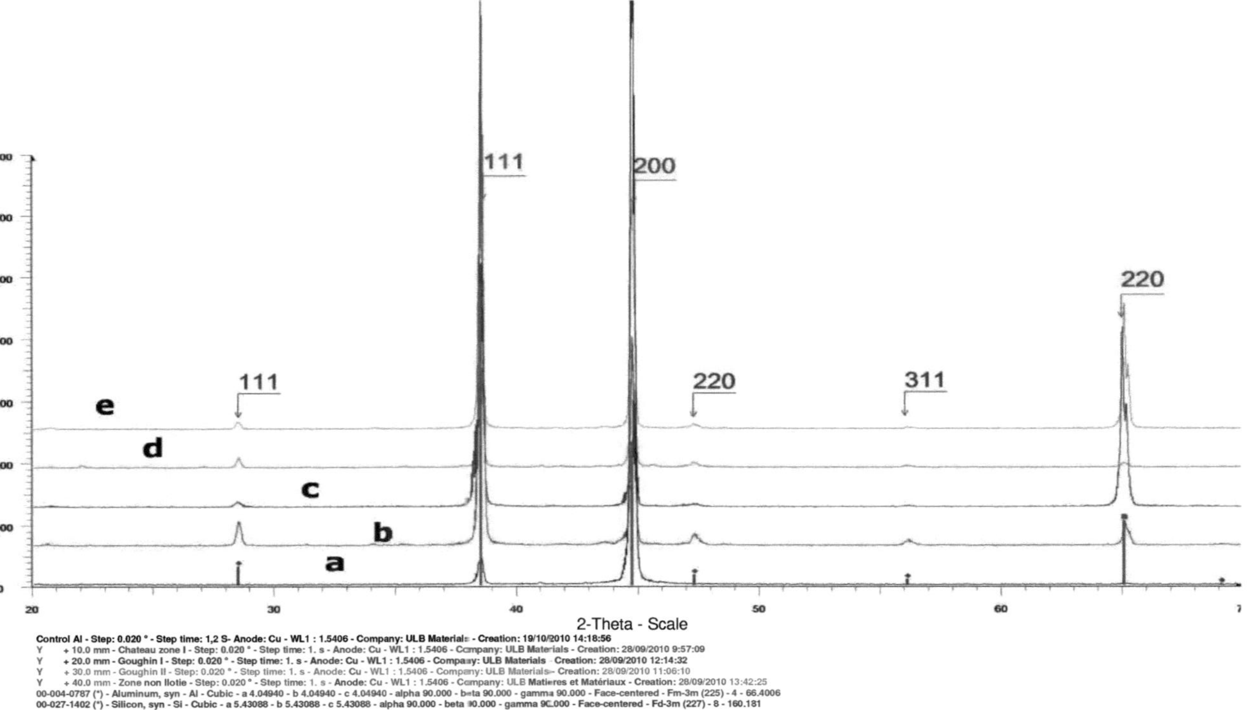

Figure 30: X-ray diffractograms (XRD): a: control sample, b: Chateau Zone I, c: Goughin I, d: Goughin II, e: Non-Lotie Zone

V. CONCLUSION

In conclusion, the acquisition of knowledge takes place during an apprenticeship lasting from a few months to several years, the terms and conditions of which vary from one region of the continent to another. There is, however, one remarkable constant in the content of the apprenticeship: each apprentice receives the totality of the knowledge necessary for the exercise of the trade, even if he or she is later called upon to specialize in a single stage of the operating chain. The apprentice who has completed his training becomes an independent vector of the foundry activity. Once established as a craftsman, he in turn passes on his knowledge to apprentices. The very functioning of the workshops implies the participation of a series of actors engaged in a training process that they often pay for. The most experienced of them generally replace the boss, who can then go on to other activities. These apprentices, who are generally referred to as "sub-bosses," also take on the training of the younger ones, in a process of "legitimate peripheral participation" as described by Ramdé, 2009. This type of transmission has an exponential character that explains the rapid spread of technical knowledge, at least at the local level.

The quality of the finished products is highly dependent on a diversified supply of waste materials, in quality and quantity, and on the empirical knowledge and experience in "materials science" of the craftsmen. This quality of the finished products is verified by the analyses and characterizations that we have carried out. Microstructurally, these materials, although very diverse in chemical composition, consist of an aluminum matrix in which we find systematically, but in variable proportions of silicon needles, intermetallic Al2Cu and Fe2Si2Al9. To this basic structure can be incorporated, according to the composition of the alloys, the elements such as Mg, Mn and Ni which one finds very imbricated in the complex aggregates formed from the two compounds quoted previously. The formation of this multitude of constituents can explain the presence of the thermal accident noted on the thermal analysis curve of the various alloys.

VI. ACKNOWLEDGEMENTS

This work was sponsored by the University Cooperation for Development (CUD), which became the Academy of Research and Higher Education (ARES) of the French Community of Belgium. We would like to thank this institution for its very important financial support which translated into research stays at the Université Libre de Bruxelles, in the framework of the institutional cooperation project between the French Community of Belgium and the University of Ouagadougou. We also express our gratitude to Professor Emeritus in Analytical Chemistry Jean Boukari LEGMA of the University Saint Thomas Aquinas (USTA) for his technical, scientific and moral support.

VII. BIBLIOGRAPHIC REFERENCES

Agier M, (1981). Foreigners, landlords and bosses. L'improvisation sociale chez les commerçants soudanais de Lomé Cahiers d'Études Africaines. **21** (81): 251 - 265.

Baralis E, Ceri S, Paraboschi S, (1995). Improved rule analysis by means of triggering and activation graphs. Rules in Database Systems. 163-181.

Baudry G, (2007). Study of the flammability of a cloud of partially oxidized aluminum particles. Université d'Orléans. pp. 285.

Elliot RA, Orowan E, Udoguchi T, Argon AS, (2004). Absence of yield points in iron on strain reversal after aging, and the Bauschinger overshoot. Mechanics of materials. **36** (11): 1143-1153.

Facy G, Pompidou M, (1983). Précis de fonderie: méthodologie, production et normalisation . : 1 p. AFNOR, toimittaja

Herbulot F, (2001). Techniques de l'Ingénieur, Traité Matériaux métalliques, récupération et recyclage de l'aluminium, strategy, France.

Müller JP, Steinegger A, Schlatter C, (1993). Contribution of aluminium from packaging materials and cooking utensils to the daily aluminium intake. Zeitschrift für Lebensmitteluntersuchung und-Forschung A. **197** (4): 332-341.

Ndiaye MB, (2006). Recycling of metals of industrial origin in Senegal. Ecole Centrale de Lyon (France). pp. 270.

Ndiaye MB, Bec S, Coquillet B, Cisse IK, (2009). Evaluation and improvement of the quality of Senegalese reinforcing steel bars produced from scrap metals. Materials & Design. **30** (3): 804-809.

Paul B, Joseph W, Nicolas M, (2006). Solid waste recycling in Fada N'Gourma

(Burkina Faso) Current practice and potentialities. :

Ramdé T, (2009). Corrosion behavior of recycled aluminum alloys for the manufacture of cooking pots. Unique thesis, University of Ouagadougou, Ouagadougou. pp. 109p.

Redjaimia Mohamed EL Hadi, (2008). Aluminium and its alloys, Thesis Mohamed Khider University Biskra, Algeria.

Sawadogo J, (2015). Physico-chemical characterization of artisanally made kitchen utensils in Burkina Faso. Unique PhD thesis, University of Ouagadougou - Burkina Faso.pp. 217 pages.

Sawadogo J, Bougouma M, Ramdé T, Boubié G, Bonou LD, Ogletree M-PD, Legma JB, (2014). Physico-chemical characterizations of cooking utensils (artisanal pots) manufactured in Burkina Faso. Journal of the West African Chemical Society. **037** 18 - 28.

Shabestari S, (2004). The effect of iron and manganese on the formation of intermetallic compounds in aluminum-silicon alloys. Materials Science and Engineering: A. **383** (2): 289-298.

Sinou ON, (1966). Metallurgical studies of aluminum objects. École nationale supérieure polytechnique de Yaoundé.pp. 98 Pages.

Tamari T, (1991). The Development of Caste Systems in West Africa1. The Journal of African History. **32** (2): 221-250.

Wang D, Shi Z, Zou L, (2003). Liquid aluminum corrosion resistance surface on steel substrate, Applied Surface Science, 214: 304-311.

Woodson G, (1998). An interesting case of osteomalacia due to antacid use associated with stainable bone aluminum in a patient with normal renal function. Bone. **22** (6): 695-698.

Youssouf A, (1996). report on pot founders in grand comore. : 9 pages

Printed by Books on Demand GmbH, Norderstedt / Germany